MW00967496

MODEL VIEW CULTURE

New writing on technology, diversity and culture.

Quarterly No. 3

San Francisco, California

September 2014

SIL Open Font Licensed Typography by:

Lato by Łukasz Dziedzic, latofonts.com

Libre Baskerville by Pablo Impallari, Rodrigo Fuenzalida, impallari.com

Abril Fatface by Veronika Burian & José Scaglione, type-together.com

Printed by 1984 in Oakland, California USA

This Quarterly is offset printed in the USA by a woman-owned small
business, on recycled paper, using zero-VOC and recycled inks, with
animal-free bookbinding.

ISBN: 978-0-9915166-2-9

Inquiries regarding requests to reprint all or part of
Model View Culture Quarterly Three can be directed to
modelviewculture@gmail.com

To order directly from the publishers, please visit:
http://modelviewculture.com

This publication is lovingly dedicated to Model View Culture's supporters - subscribers, community sponsors, and patrons. You make it possible for us to remain independent, and keep publishing critical and diverse work.

Thank you for all you do. <3

Contents

Letter from the Editor

Welcome to the third issue of Model View Culture's 2014 subscription!!!

Did you know most of our operating budget comes from you, our subscribers? Thank you, so, so much.

Here's a quick update on us!

Since we last spoke, we've been pretty busy. Over the summer, we published online issues on management, social media, open source and mental illness and disability. By the time you read this, our issue on sex and sexuality should be online too.

To continue to build the sustainability of our business, we've rolled out individual sponsorships for our online issues, as well as a Patreon page where readers can give us a small amount of money for each one we make. We're excited to have new and different ways readers can support us.

We're starting the process of making some design changes to our website, modelviewculture.com, so that the site continues to grow and evolve as we do! As our archive of issues and articles grows, one of our major goals is to make sure that content is discoverable now and in the future.

As we head into Autumn, we're thinking about our editorial calendar for the rest of the year, the last quarterly of the year (you'll get that in a few months), and what lies ahead in 2015.

It's been an incredibly exciting first year so far, and we can't wait to see what the rest of it holds.

A few thoughts on Quarterly 3:

While most of our online issues have a specific theme, our Quarterlies give us the space to explore lots of different topics in one collection. We get to meet and publish new authors, explore new things we're thinking about, and put it all together into a unique and exclusive compilation.

It's a challenging and rewarding experience to bring people together from all different sectors of tech, to explore important topics in our field. In this one, we cover a lot of ground: publishing, surveillance, management, hierarchy, gamification, online activism and more.

We think you'll find this issue to be both challenging and hopeful, wide-ranging yet rigorous. And as always -- diverse, critical... and uncompromising.

We didn't have much room for this note with so much great writing to fit into the pages. We hope you'll enjoy reading the issue, and then pass it along to someone else.

Shanley Kane, Editor and CEO

Interview With Julie Ann Horvath

*Our editor sat down to talk with Julie Ann Horvath, a design-
er and programmer at &yet. Julie is well-known for her work
starting Passion Projects, a platform which features talks from
women doing amazing things across the tech industry. Earlier
this year, she went public with her experiences working at
GitHub, in the process changing the industry conversation on
tech culture. In this interview, we talk about accountability for
tech leaders, speaking out and making a difference.*

**Well let's just get started. Something you've talked about
a bit is the tech community developing accountability,
and how that feels like a major change that's happening
right now in the industry.**

That's the big part about social media activism that ex-
cites me. It allows people to have a voice instead of people who are just in the boys' club, or are "tech illuminati"
or something like that. There are aspects of speaking out
that are born out of privilege -- I had already found a
new job at the point I spoke out about GitHub, so I was
able to voice my story. But I do like the idea of the community forcing our leaders to be accountable for their
actions - for me, that's the most exciting thing going on
in tech.

You look at the last month or so, how many terrible
people have been removed from positions of power in
the industry for abusing those positions? Like, I'm sure
most of those people are still very happy on their beds of

money, they still have comical wealth, but they are facing some sort of accountability for their actions.

Speaking out is so important. At GitHub, they tried to get me to sign a non-disparagement, but for me the only thing I've ever had is my truth and my story. I live really well now because of my job, and I'm good at my job, but as far as giving up my truth, the price on that is huge for me.

They were trying to get me to sign a non-disparagement agreement and they were like, give us terms under which you would sign this. I said I wanted 15 million dollars and Tom Preston-Werner's resignation. We had a good laugh, and obviously the lawyer could not negotiate those terms on my behalf (shocking). But I was just happy to have the founders hear that that was how much my experience was worth to me, and being able to share that experience.

The non-disparagement agreement issue is such a big deal. I don't think people realize how many times a non-disparagement agreement is brought in right away when there is wrong-doing that's happened and discrimination and inequality is involved.

I think a lot of marginalized people who have had really bad experiences in tech, need that non-disparagement money to survive. They end up feeling forced because of their circumstances or the circumstances of their family to sign when leaving a company, and say OK, I'm giving up my truth and my ability to tell my story.

I think a lot of those stories would have helped other people who are now coming into tech, because you have to be able to know what you're getting into. Because otherwise you wake up one day and you're like "Holy shit, I've been complicit in all these behaviors and actions," because you didn't realize how many people they were hurting or you just didn't realize that it's not normal. It's easy to feel like the tech culture is just normal and how the industry is or should be. And that's the moment of maturity - going from "oh, this is how it is" to feeling like no, this is wrong and this isn't how it should be.

When you're starting out, all you want is to do well in your career. You think, I have to be able to be "one of the boys" and be able to hang with these people in order to prove I'm good enough.

You have to reach the point of realizing that that's not good enough, and that they will still hurt you. That one day, you'll find out that no matter how much a part of the boys' club you've become, you are not one of them. For me I reached that point at GitHub. It's also really powerful to realize you don't *want* to be one of them, you don't want to live like a hamster in a really fancy hamster cage. No offense to hamsters, they're great but I just feel like, I went to high school. We went to high school, we did that.

And it was shitty then, and it's shitty now.

It is shitty. And they throw all these fancy perks at you to make you think that you're winning. You think that you're winning, you think you're part of this club, but it's funny because the moment you raise your hand and say *I think this thing is wrong*, you're immediately excommunicated from that group, you're the other, you other yourself. You're almost being paid to just be quiet and let the people who are in the really influential roles drive the company culture and drive the community and the industry, even though they shouldn't be. That's my biggest problem with all of these really young people who don't have experience leading companies and that whole thing, is that I don't think the people who are in leadership positions should be.

When we talk about management in the Valley, often managers are white men who have no qualifications.

Absolutely not, they're good at whatever it is they do, a very small spectrum of technical skills or part of the thing or product that they own. But that doesn't make someone a good manager. Oftentimes you get people who are very volatile, don't have a lot of empathy for other people, or know how to talk to other people, or understand what they're going through, their reality.

When I was going through my situation, I felt so cornered. I would have done anything for people to just be happy with me, because for me losing my job affects my family back home and I help them out a lot. That's one thing that's great about tech, having a high-paid salary allows me to do that, but at the same time that makes the idea of losing my job so much more threatening and scary.

And when you do speak out, you just get accused of being manipulative, or trying to profit off the situation.

I've had a lot of issues with the conversations that happen around people who date at work. The idea that if a man aligns himself with a woman in any way, and not even in a relationship, but if he supports a woman's voice, she must be tricking him or manipulating him with sex. Like, there is no other way for us to be. That's how people suggest that we come to power, or become more visible, is that we're manipulating people or tricking them with sex. When really we're not asking for the men in tech to project their nerd girl fantasies on us, they're doing that all on their own. We're just being ourselves and doing our jobs.

When your goal is to succeed in the industry for so long and you finally do good, but you're being treated like shit, it's still so hard to walk away because it's like, *this is what I've been working for.*

Yeah, feeling that *I deserve to be here.* I started going and giving talks, I was posting a lot internally, but it's funny because when I do that as a woman it's construed as self promotion, it's being cocky or having an ego. But when men around me are doing that, it's normal or what they are supposed to do. It's fucking ridiculous to me that I didn't even do those things as much as men at my company, who were rewarded for it.

I thought that's how I become a good GitHubber -- that's the idea of success. Especially if you don't have a management team who is defining what success means in your specific roles, you're going to build your own idea of success based on what you see around you. And if you're a woman or a minority or you're not in that inner circle, you're going to be torn down for it. Punished, even.

We don't have models of what success for us looks like because all of the bros around us who are succeeding are doing x, y, z, but if we do x, y, z, we're fucked.

This is funny because this happened on pull requests all the time and it happens in open source too. You get responses to your work disguised as feedback that are just brutal attacks. When I matched that tone or gave it back to them, *I* was being too aggressive, or *I* was out of line. Literal quote, "out of line," for giving them back what they're serving or voicing any concerns, personal or otherwise.

Another thing that gets lost in this conversation about media activism is people think we're telling these horrible salacious stories, overdramatizing them, and really they're hearing 1% of what happens.

You almost can't say the rest, because if you say anything more, then you're doing it for attention or it becomes this thing, a broken record. I was talking to my friend who works at a tech company, he even knows that if you raise awareness about one issue, and you want to keep your job, you can do that like once in a year. Otherwise it becomes less credible, and that's fucked up because we're only allowed to share a little bit of what we've been through. And financially, if we do say it all, companies have hundreds of millions of dollars and whole teams of lawyers, unlimited resources to completely discredit you and ruin your career and ultimately, your life.

If they took any of us to court, that's not only potentially a lot of money if we're being sued for something like libel, but it's also a lot of our time. And everything about your life would then become public record. I don't think people understand how much we're risking by telling our stories, and then because of that, how few women actually do.

It's funny, women will actually shame other women for coming out with their stories because they feel like *they* can't do the same or that it's starting "drama". The worst and arguably the most shocking behavior I encountered after coming out with my experience was from women I worked with, women I called my friends, who I both supported and advocated for internally at GitHub. Women are always the first to discredit other women, likely because they've personally experienced something worse and didn't or couldn't speak up for themselves. But to tell you the truth I don't know what's worse, the attempts to discredit me or the complete silence from women who had voiced similar concerns about the culture at GitHub long before I chose to leave it. When thinking seriously about whether or not to disclose conversations like these publicly, I ended up choosing not. I guess at the end of the day, I still believe in protecting other women from the abuse I've endured. I don't want to punish anyone. I'm not sure the feeling is mutual, but hey, I learned something. *This* was by far the most eye-opening part of my experience.

Another thing I wanted to talk about is HR. Now more and more people are talking about HR, debating the roles of HR at startups. In your mind, what do you think companies should be doing about HR?

HR should be a really early hire. It should be someone with experience in HR, it should be someone who can be trusted, you should be able to tell them things in confidence and know they are being worked on. It's funny because as minorities, we're always told we're the liability or the risk, but to be honest, a cofounder that is emotionally torturing one of your employees, a situation that could potentially turn into a very public and humiliating thing for your company... THAT'S the risk. That cofounder is the risk. That person needs to be removed or at the very least held accountable. Who is going to do that if there's not someone in an HR position that has a significant amount of influence in that role, that is a representative for those risks?

Here's the thing, if you needed an engineer to build a feature that the company really needed, you would never hear them be like "oh, we searched for a year and couldn't find one." It wouldn't happen. But these companies say that all the time to explain why they don't have HR.

Another example of negligence. A lot of these startups are just purely negligent and they're not being held accountable for that whatsoever. HR is a good example, another example is tax law. They think they're pushing laws to be innovative but in actuality they're just negligent and they're hurting lot of people around them. The game is rigged. If you don't have access to unlimited resources, you have to play by the rules. As soon as you have access to those resources, you can just do whatever you want and call it innovation or "disruption."

Sadly even though there is more public awareness, many of the people who are building abusive work environments will be OK while whistle blowers face serious consequences like not being able to work in the industry anymore.

They get handed the opportunity to lead another organization and make the same mistakes again because they're not being held accountable for them. But yet, as people who speak up about it, *we're* the bad guys. Like *what*. That makes absolutely no sense.

You started an awesome program for women in tech, Passion Projects. Do you want to talk a little about that and what would be your advice for people who are looking to start projects in that space?

My advice would be, find people who have a mutual interest in starting these things, and build your team out of that. One thing I felt really lucky to have was immediate support. Just putting something out there, people who care about what you're doing gravitate towards you. Working on something with those people is so much better than trying to do it properly through organizational channels. Just say you're going to fucking do it.

I got on stage at the GitHub employee's summit. You had 5 minutes to give a lightning talk on something you care about. I had been working on Passion Projects kinda secretly until I was ready to ask for what I needed to make it happened. As my life was a living hell at GitHub, that was brutal, but I was like "I'm going to do this thing because I *have* to do this thing."

It just so happened the only founder in the room at the time I was presenting was Tom. I was like well, if he's going to say no to me he's going to have to explain to 200 people in the room why he said no to this thing, which is going to largely benefit someone who was not him. And you know, he just gave the thumbs up.

My best advice is just put your foot down and just fucking do things until someone tells you you can't, because I guarantee it if you're doing it for a cause you believe in, and something that's really important - like raising awareness and serving the work of awesome women and minorities in tech - no one's gonna get in your way, because the moment they do, they become the enemy. And how different would it be for them to be on the other side of this? For them to have to be that person and walk in our shoes.

The thing is they wouldn't be able to go through what we go through. They would lose their shit. Try running a company, doing your work everyday, with non-stop harassment every day. Good luck.

They couldn't do it. They could not do it.

Engineering Management and Diversity

Marco Rogers

I've been an engineering manager at Yammer for a year and half now. This isn't my first rodeo, but my previous experience leading people was over 5 years ago. These two experiences have been very different, and the gap between has given me a lot of perspective. At the core of this perspective is a recognition of two different kinds of "management".

The first type of management I call command-based. This is the one most of us assume when we hear the word "management." Here, the manager is the top contributor, and treats the team as an extension of themselves. This makes sense to most people, because managers are usually chosen based on their success as an individual contributor. Their job is to get a team of people to do what *they* would've done.

In this model, engineering managers are primarily responsible for the success of projects and the quality of the work. The team is expected to follow their lead and defer to their judgment on important implementation decisions. In many orgs, the manager is even responsible for breaking down all tasks and assigning them to individuals on the team.

But this approach has some consequences that are not obvious.

Managers in a command-based org are under a lot of pressure. They take on all of the responsibility for the outcomes produced by their team. Engineering managers are held accountable for bad code, failing builds, production outages, etc. They may lead a large team of engineers, and they may spread that accountability around by exerting downward pressure. But eventually, the buck stops with them.

We all know it's impossible to have full control over outcomes. Yet a manager in a command-based org is asked to control the outcomes of a group of engineers using any means at their disposal. When the team faces challenges, you see this control exerted in many ways. If you've ever had this kind of manager, you're probably familiar with these. We talk about "micromanagement", where the boss is constantly looking over your shoulder and questioning your work. We see strict timekeeping where managers want to be able to account for the hours everyone spends working. In engineering, this is at least partially responsible for managers who also serve as "architects". They control the work by describing the system in detail and demanding that it meet their exact specifications.

At least in software, where I'm most familiar, there's the sense that engineers will go off the rails if left to their own devices. The manager's job is to look over their shoulders, keep them on task, find missteps before they get too big. This fits well with a command-based structure. The trouble is that the longer the manager stays away from producing work themselves, the less their judgment should be trusted by the team. This is the core problem with manager-as-top-contributor: the longer you are a manager, the less likely that you're able to maintain your level of skill and competency at the work the team does.

Command-Based Management and How It Impacts Diversity

It's not difficult to see many of the problems caused by command-based management. But when we focus on diversity, we can uncover a more subtle effect. The tech industry has come under fire in recent years for its abysmal track record on diverse hiring. This industry is overwhelmingly populated with white males with the only other significant ethnicity being Asian males. Women, minorities, and other marginalized groups are woefully underrepresented. As we go up the ladder to management, these numbers get even worse. I believe the perverse incentives of command-based management can help us understand at least some of these factors.

As command-based managers are accountable for all outcomes of the team they lead, they're incentivized to find ways to control as many factors as they can that affect these outcomes. I've mentioned some of the obvious ones. But another important factor in team success, arguably the most crucial, is hiring the right people. A manager needs to hire the people they think are most likely to do things the way they would've done. And it's here that we find a great deal of bias.

How can a manager know if a candidate can produce the way they would? We learn how to evaluate technical skills, but that still leaves so much to chance. Do they work hard? Are they reliable? Will they work well with the rest of the team? It's difficult to get confident answers to these questions before hiring. But it's easy to fall back on our own experience. It's easy to take a liking to that person who has a similar background to us. Maybe they went to the same school. Maybe they started with the same programming language. Maybe they just remind you of yourself, or other smart people you've seen be successful.

Most of us would never admit that these factors affect our hiring decisions, though sometimes people acknowledge this thinking -- like when Paul Graham admitted his bias towards founders that look like Mark Zuckerberg. These biases make it very difficult to take a chance on people you can't relate to. If you're an engineering manager, you're probably used to engineers who look a certain way, act a certain way. It's tough not to fall victim to the norms that have developed around tech culture. If all you ever see is white men doing this job, how will you react when it's a woman sitting across from you in an interview? Can you really see a black or Latino person being that awesome geek you pictured yourself hiring?

In tech, there is plenty of evidence that says the answer is no.

Service-Based Management

Command-based management has lots of flaws when it comes to building functional and diverse teams. But there's another type of management: one where the manager isn't responsible for the work, but for building a team that is effective at execution. The manager's job is to support the team and facilitate whatever they need to be successful. I call this service-based management.

Many managers read this and think that's the job they do today. But there are some key differences. Most notable is that the service-based manager is not seen as the top contributor. Maybe they used to be when they were on the team. But when they accept the new role, their focus needs to shift. The manager is going to produce very little work, maybe even none at all. The responsibility to consistently produce quality work falls squarely to the team.

A service-based manager doesn't participate directly in deciding how the work gets done. The manager's job is to establish high expectations for the team and set direction to help the team to meet those expectations. They establish trust with the team, understand and remove obstacles, and help the team develop ownership over their work.

A service-based manager is still intensely concerned with team outcomes. And outcomes can still be measured even if the manager was removed from the process. In engineering, some of the same tools apply. Monitoring the success of releases, number of bugs generated, etc. If these outcomes don't look good, you take this information to the team: "Our bug queue is out of control. What's up?" This is a test for the manager and for the team. Does the team know what the problem is? Can they communicate it effectively to the manager? Does the manager trust their judgment?

In a command-based system, it's questionable whether these things are true. The team expects the manager to be on top of things. This sometimes means they have a narrow view of the world and don't always have perspective on the problems that occur. A service-based manager has already set the expectation that *the team* owns the bug queue. So they should know what's going on with it. They should tell you immediately why it's in the current state and what they're doing to fix it. If they can't, then they are failing to meet expectations and that has to change. The next step is giving the team the space to fix the problem. One of the major themes of service-based management is ownership. The team will excel because they know they own the problem. They don't expect anyone else to do it.

Creating a cohesive team and building trust is part of a service-based manager's job. It takes a lot of time and a lot of empathy to get the team to trust in their manager. My experience with 1-on-1 meetings with command-based managers haven't been very useful, because it's usually about control. "Here are your tasks. How are you doing on your tasks? Let's talk about some tasks you didn't do." A service-based manager trusts that the individual in front of them knows what to do. They are looking for ways they can help.

My 1-on-1s now consist of things like, "How's this going? Any roadblocks? I can take care of that". It's also a time for constructive feedback, both positive and negative. "It looks like you're having trouble with this. Let's talk about how to improve." Building trust is a constant conversation.

Service-Based Management and Diversity

Taking a service-based approach to management can also help with some of the challenges of diverse hiring. Changing the focus to *serving* the team provides opportunities to avoid bias.

In a service-based system, the manager isn't directly responsible for the work that gets done. This has the effect of removing the problematic incentive to exert tight control over *how* it gets done. Micromanagement of individuals on the team is no longer necessary when outcomes are more important than controlling the work. In fact, giving people more freedom and autonomy in choosing how they get their job done has been shown to make them more productive.

When you value autonomy and flexibility as a manager, you can create an environment that is inclusive of different work styles. People who get tons of work done during off-hours have less pressure to also be productive from 9 to 5. People with families can leave early or come in late when they need to, without the stigma that they don't work as much as the rest of the team. A team that focuses on outcomes can welcome people with lots of different circumstances and backgrounds rather than matching a certain "culture fit".

Having a manager shift focus to building a productive team also has benefits when thinking about hiring. A command-based manager might ask "how can I find a great engineer to fill this position"? This person is likely to fall back on those norms of what a great engineer looks like. But a service-based manager instead goes to the team and asks "what kind of person would make this team more successful"? Not only does it leave open a wider range of answers, it helps the team have a more nuanced view of candidates.

Involving the team in hiring can help avoid a monoculture when they look to add strong compliments to the team across a wide range of skills: "I think we would really benefit from this person's expertise in web accessibility." And a cohesive team will expect to work together to support each other, rather than judging people where they may be weak: "They stumbled a little on algorithms, but I think they can rise to that challenge with some experience."

The hiring focus in this model is evaluating a wide range of skills and qualities in a candidate, and bringing the whole team's perspective to bear on hiring. It becomes much more difficult to judge candidates just based on good coding skills and a gut feeling. The team is incentivized to create a more rigorous system to gain confidence in new members. And the manager can help with that.

These are only a few aspects of how outdated organizational systems can make it more difficult to create the kind of environment people can thrive in. Moving away from the traditional notion of command-based management is a great step towards companies that can be more inclusive as well as more productive. Exploring different approaches to management can have profound effects on your team and your organization.

Detroit 2.0

We sat down to talk to Paul Riser, Jr., the new managing director of technology-based entrepreneurship at TechTown Detroit -- Detroit's most established business accelerator and incubator.

Do you want to tell us a bit about the tech scene in Detroit right now?

Immensely vibrant, very energetic, very lively, very optimistic. We find ourselves now at a tipping point of great investment, great entrepreneurs, great engineering capital, and great intellectual capital. Building off our rich history in automotive and engineering, there's a sense of a new, Detroit 2.0 perspective. It's just a dynamic place right now, with huge activity. Everything from conferences to hackathons, seminars and meetups, and startup accelerators, incubators and coworking spaces. There's also a huge uptake in optimism and belief from the entrepreneurial community of what can be, and what will be.

We also have great buy-in from the mayor, who has announced a new innovation district. It's a huge effort to merge what's going on in our Downtown and Midtown corridors, creating a dense level of entrepreneurial activity, storefronts and small businesses. It's also about pushing that momentum, the best practices and knowledge out to Detroit neighborhoods to support individuals, small businesses and the entrepreneurs who might not ever make it to TechTown's doorstep, or might not ever make it downtown to seek entrepreneurial support services.

Can you tell us a little about TechTown Detroit and what you do there?

Under the president of Wayne State University in 1999, the leadership identified an unmet need for business acceleration and incubation that could bring the University's extensive pool of expertise and business insight to bear on diversifying the city's economy. By 2004, the idea got legs -- a number of other instrumental partners jumped on board and created what we know now as TechTown. TechTown itself is just a truly dynamic place where technology, talent and capital converge. It serves as a catalyst for an entire community that spans across Detroit, for startups across retail, wholesale, medical device, advanced manufacturing, and other technology sectors. We continuously host coworking, incubation and other events that support all of that catalyzed activity. So even though we're an independent nonprofit, we remain a staple in the community for strong relationships and entrepreneurship.

We're a resource not just for people who are in Midtown and Downtown -- those areas are well-supported and have great anchors from an innovation cluster perspective already. But in order for the entire city to thrive, we have to have inclusion for those who are even on the edges of Midtown and Downtown, who are still stakeholders and citizens of this great city. We want everyone to build as if they have a participating role in this new Detroit 2.0.

What does diversity in tech look like in Detroit, and what are the issues and opportunities?

Diversity is another indicator in Detroit that's on the uptick. But, we also face our fair share of challenges with the level of inclusion. Detroit and Southeastern Michigan are some of the most diverse regions in the world from an ethnicity and background perspective, representing dozens and dozens of ethnic backgrounds and ethnic cultures. This diversity is a huge asset as we expand and really redefine the city's future. I think it's imperative, and it's unquestionable, the value that diversity brings. We must ensure we reach out to African American communities, Asian American communities, Arab American communities, and on and on, to make sure we have a diversity of not just complexion but ideas, concepts, talents and resources that are part of those rich communities.

As far as the investment landscape, what does that look like?

We have had the opportunity to learn from what has occurred in other places, what has worked, what has not, and really define what *we* want to be. We have a great blend of what has occurred in the Valley and what has occurred in the East Coast and Boston areas. We are really respectful of tradition but also focused on novelty. So we're looking at some of those emerging technology fields -- telemedicine, telehealth, mobility. And with us having the automotive industry, we have this great industry of interactive vehicles, vehicles that not only communicate but autonomous driving vehicles. The biomedical space and life sciences space is a great area because of our rich culture on the academia side. We have a great culture of research-led initiatives and projects that TechTown is involved in to help spin out great technologies.

We're attracting outside investment more and more, and we have more and more venture capital and angel funders who are creating a presence in the state of Michigan and Southeast Michigan particularly. We're getting more support at the state and municipal levels as well.

You also run a magazine dedicated to black arts, life and culture, which is so cool.

I am an equity partner and owner of BLAC magazine, which has served the Southeast Michigan community for about 15 years. I got in when it was about 7 years old. I saw it as an opportunity to really grow it into something that would serve not only the African American community, but be a model for positive news and positive reflections of life, arts and culture in a particular demographic.

Sometimes our media sources can tend to beat up on diverse communities. In Detroit we have been victims of our own bad news in some instances. And so even in our most challenging times, BLAC has always found positivity, relative to people, to opportunity, to culture that has always been a shining light in our community and kept that sense of hope for people.

What's your advice and thoughts on what Detroit has learned over the past few years that is applicable to other growing tech hubs across America and even the world?

It's critical to engage the private sector and industry, the academic community, your health centers. These are strategic partners that you can leverage in order to create the density of attraction and retention when building an emerging tech community.

It's not an easy thing to reimagine a city's identity and business models, to create urban sustainability and connected technology, without inclusion. Many cities are grappling with the complex challenges of stimulating economic opportunity, adapting to various changes -- such as urban liveability, green transportation and energy, finding ways to be more resilient. Inclusion is so important.

Gamification, Abuse and Activism

Betsy Haibel

No one likes to think about 4chan. It's the Internet's id: an inchoate howling forest of proto-memes and wannabe black-hat hackers. But 4chan *works*. 4chan's creator, Christopher Poole — better known as moot — defends the system he built as a well-oiled creativity machine. He's right, though that creativity often takes ugly forms.

4chan works because of the community's cutthroat nature. Poole talks a good game about the liberatory power of anonymity — that it allows 4chan's "dialogue" to be "as honest as it gets;" that this honesty, combined with an enforced short attention span, has bred an unmatched "meme factory." He's not wrong, but his analysis elides 4chan's reputation as an unsafe space — the crucial last part of its formula. 4chan's ambiently abusive norms motivate its denizens' internal editors. If you produce something that amuses 4chan's collective, um, soul, you're rewarded. If you don't, you become one of its targets.

4chan's community is an emergent reward-loop engine. Reward loops are central to gamification — the practice of incorporating game mechanics into non-game software. More bluntly, gamification is the practice of using stimulus-response engineering to encourage specific user behaviors. Shallow implementations of gamification tend to focus on cliched, surface-level mechanics: things like badges, points, and leaderboards. These shallow manifestations are easy to duplicate, but they miss gamification's true and effective heart: creating a genuinely compelling micro-reward, and figuring out a way to give it to users that maximizes a desirable user behavior. In 4chan's case, the micro-reward is a hit of social approbation; it's granted for the desirable user behavior of producing interesting content.

This reward engine differs from ordinary human status-seeking because it's on a computer. That's not a cheat — gamifying social behavior is possible because social-media behaviors sit at an uneasy intersection between real and unreal life. It's overly simplistic to suggest that what happens on the Internet is not real. The Internet, nowadays, is a place where people get jobs, learn about parties, crack jokes, and fall in love — and online abuse creates very real cases of post-traumatic stress disorder. That said, when people interact on the Internet there is a genuine layer of often-dehumanizing unreality interposed between them. This mediating layer is consciously designed, and it's where the manipulative aspects of social-media UX come into play.

Likes, retweets, favorites, and upvotes are all point systems. The interesting thing about point systems isn't the points themselves — numbers are abstract — but rather, the thing that the points represent. Humans tend to seek social approval, and social-media point systems *quantify* that approval. Suddenly, it's possible to measure minute gradations in how compelling a statement was. This, in turn, motivates people to beat their scores and the scores of others.

Social media offers other point systems as well — follower counts, @mentions, Facebook share counts, and other aggregate "social reach" metrics. These can serve as longer-running overall assessments of player performance, an attitude reinforced by companies like Klout which have parasitically latched on to the social-media ecosystem.

Social point systems are morally neutral. The UX goal behind them is typically "increase engagement;" this goal is agnostic about that engagement's semantic content. Since social point systems are social, they reward people for posting content that's in line with community norms, and as such amplify those community norms. When those community norms are those of 4chan, people optimize for lulz.

Real Life If It's Hurting You

Online abuse takes forms nearly as varied as online porn. There's the Twitter death threat, the email death threat, the DDOS against one's employer. There's the cyberbullying that high schoolers engage in, peer-to-peer. There's revenge pornography, where old sexts are uploaded as an act of aggression.

Many forms of online abuse are inherently performative. All can be *made* performative when a harasser later recounts their actions, as many do. And all performative forms of online abuse are potentially subject to the amplification mechanisms baked into social media by gamification. Additionally, some online harassers may get their jollies from the act of harassment itself. It feels like cheating to class that as a gamified behavior, but it is technically a desirable user behavior (content generation) that is made easy and that the user is rewarded for performing. Nothing about gamification rules out the additional presence of intrinsic motivation.

Online harassers may be broadly broken down into two categories: those who are engaging in online harassment because they want to harass their target and it's an available means of doing so, and those who are engaging in online harassment because they're bored and it's fun. Simple electronic deterrent measures, like blocking, tend to not have a significant effect on the behavior of motivated harassers but can often bore the unmotivated into either going away or finding a new target. Gamification measures provide some positive reinforcement for motivated harassers, but they'd harass anyway. Their effect on unmotivated harassers is far more marked — gamification measures largely create this group by providing enough positive reinforcement to make harassment worth the work of harassing.

Many motivated harassers view opportunistic harassers as their own private troll army. They choose to efficiently scale their harassment campaigns by recruiting foot soldiers through doxxing campaigns or sometimes just their own public mockery. They only need to spend minimal effort on this recruitment process because so much of the work of motivating opportunistic trolls is done for them by various social sites' gamified content engines.

Apoliticality

While vicious attacks by groups such as 4chan comprise some of the most visible cases of online harassment, this harassment is merely the well-publicized tip of a vast, political iceberg. It's easiest to see the political connotations of harassing activists - after all, documenting and articulating one's personal experiences is a major tool in the online-social-justice toolkit. But very little online harassment is apolitical. The gruesome misogyny of revenge porn is pretty obvious. Teenage cyberbullying — like teenage bullying throughout the ages — tends to be directed at targets less privileged than the perpetrators (queer teens being bullied by straight ones, working-class teens being bullied by upper-middle-class ones) or is used to reinforce gender roles and stereotypes (slut-shaming).

Some groups — notably women of color on Twitter — are subject to organized harassment for daring to exist in public, whether their tweets are overtly political or not.

These campaigns, such as the one exposed by #yourslip-isshowing, are often framed by their organizers as activism-by-harassment.

In all these cases, the "unreal" game layer of online social interaction serves to amplify the abusive aspects (because perpetrators can turn off their empathy; because UIs push them towards engagement with a system) but is not terribly effective in reducing the harm done. The Internet is real life to the people hurt by online abuse, and not real life to the people doing the hurting.

Activist Hypervisibility, and Entitlement to Consumption

Online abuse also occurs in the everyday contact and communication between users of a platform. Twitter, like many social media platforms, flattens out relationships between users and the people they follow or interact with. While there's a world of social difference between tweeting at Nathan Fillion and tweeting at your ex-roommate, there isn't any programmatic difference. This is by design. Twitter's appeal partly lies in the way it facilitates its users' parasocial relationships with celebrities and other interesting strangers.

Many fair-weather allies are partly in the game for access to these parasocial relationships. Because social media frames parasocial relationships as ones with celebrities, activists are turned into celebrities — often nonconsensually. Gradient Lair's Trudy has written extensively on the negative effects of this on the activists who are made hypervisible — it opens them up to plagiarism, harassment, and the oppressive gaze of those who view them as content-producing machines.

In other words, Twitter-only activism from members of relatively privileged groups sometimes has positive effects but may also have negative effects on the activists that these allies are "signal-boosting." These negative effects are discounted, however, because social-media faux-celebrity — with its attendant "high scores" in RT count — is viewed as *winning* the game that these platforms encourage. If visibility is the goal, hypervisibility is a coveted success. When the hypervisible complain of abuse, they're viewed as whiners, as people who simultaneously want the privacy of obscurity and the privilege of public-figure status. (This is partly why the harassment faced by the "winners" on the Gittip receivers' leaderboard was not taken seriously until the Gittip Crisis.) The assumption that everyone *wants* to "win" — to be a public figure — runs so deep that abuse of the hypervisible is assumed to be inherently compensated for by the "privilege" of their hypervisibility.

Social media platforms which sell "celebrities," such as Twitter and Instagram, depend on the existence of these celebrities and as such are designed to create them. Activist hypervisibility is a designed effect of social media gamification's signal-amplifying mechanisms.

Parasocial relationships are a bonus for activist *movements* in many ways, despite the pressure they place on their targets — they provide emotional attachment points for the people that activists attempt to influence. It's nonetheless reasonable to ask whether "engagement" within these parasocial relationships, and other rewards specific to online action, serve as a pressure valve on activist and ally energy and bleed off time and attention that could otherwise go to more effective work.

The Indirection Layer

Online social behavior is deeply shaped by the platforms on which it takes place — from the rich hypertext conversations of Livejournal past, to the polyphonic yells of Twitter hashtags, to the family arguments of Facebook.

The intent of this shaping is near-always towards the goal of "increased engagement" with a platform — Twitter cares a lot less about the semantic content of tweets than it does about the advertising eyeballs associated with site engagement. "Increased engagement" has at-best mixed consequences for activists and marginalized people on web platforms, and the platforms themselves have consistently chosen to ignore the moral implications of their effect on user behavior.

In Twitter's case, this ignorance is profitable and likely deliberate. Social media platforms which sell parasocial relationships, which sell "engagement" with "celebrities," have no incentive to discourage this "engagement" - even when that's an euphemism for abuse. We maintain the polite fiction that these platforms are "natural" — that our behavior on them is not shaped by invisible hands — at our own expense. They are designed to throw accelerants on fires. They are designed to sacrifice our safety for small profits.

They are designed to amplify *everything*, both our voices and our screams.

The Angry Others

Divya Manian

One thing you notice as a foreigner is how quickly you are marked as The Other. You notice it even more as a woman in tech.

I was having a conversation with a female friend at Sightglass, the VC/startup watering hole in SF, when someone was quick to remark on our status as The Others- 'You are women! You code! You're Unicorns!'

In a workplace, this unspeakable Otherness manifests itself in shared attitudes that become codified as 'work culture'. We work with people every day - holding conversations, participating in shared activities, and even celebrating progress that has been made. Sometimes, we participate in events organized at our workplaces. It is such actions, conversations, and shared common ideals that become the culture of our workplace.

Most of the time, work culture is actively built by a team within our workplace, and our daily conversations and actions inform them too. Unfortunately, the culture that exists in American workplaces is also deeply biased towards white males. Recent disclosures from Google, Facebook, LinkedIn and Yahoo validate this fact - the tech world is largely white-male dominated. In this environment, it's unsurprising that male-bonding activities are encouraged to develop team culture: drinking beer while engaging in some activity (like 'hackathons'), talking about or viewing a male-dominated sports event, or competing with another team.

People who don't belong to the groups that enjoy such bonding activities (lets call them The Others for brevity) sometimes get hired, decline to participate in them and are marked as 'not a team player.' Perhaps the team loves intense Crossfit sessions. If The Other displays no interest in such an activity, the team judges that person for lack of interest.

Not only is the lack of interest noted, but The Other soon realizes that a lot of conversations that inform success at work take place during such activities. Not participating could mean missing feedback that contributes to promotions, or just being left out of the bonding and intimacy that develops during these activities. When the dreaded re-organization happens, The Other find themselves without a team or facing a layoff because they did not become familiar enough to the rest of the team.

The Consequences of Being "Other"

When that occurs, the team decides to never hire someone 'different' again. In future interviews, the team unconsciously look for indicators of 'The Other' from past experiences - like race, gender, qualifications or even the accent of speech - to weed these candidates out.

Meanwhile, The Other considers their choices. Being flagged as The Other has consequences that are not worth sustaining. The Other either goes to work where they are no longer deemed as such, or finds a way to get rid of the Otherness. We try to adopt the accents of our coworkers, wear clothes that would not make us stand out (hoodies, t-shirts and jeans), watch TV shows that our coworkers watch, try to enjoy the music that our coworkers appreciate. This also means speaking in a tone that coworkers *expect us* to speak in - which might not be what we as Others are comfortable with.

As a woman, this is even more striking. Women are expected to have 'manners' but men are excused for lacking them because of 'genes' and the myth that they are naturally more belligerent. Women are expected to be 'kind' and 'motherly' because of the persistent myth that 'nature intended them to be that way'. This means, as a woman in a male-dominated workplace, you constantly work on how you respond in conversation to fit into the expected behaviour and thus be successful.

To behave contrary to expectations is dangerous because it can get you fired - which can have serious consequences. US companies often have employees agree to a same-day notice for dismissal, and the expectation is you gather your belongings and leave immediately when given one. If you are on an employer-sponsored visa, the risk is higher as you are expected to leave the country upon dismissal - unless you immediately find another company to employ and sponsor you. It was also surprising to me that sometimes the only way to actually get a doctor to see you without you having to declare bankruptcy is to find an employer who is generous with their health insurance benefits. This means as a woman you either have to suck it up and behave, or find a team that is mostly women to work with.

For women in tech, there are far fewer workplaces or teams that are either female-majority or diverse enough to feel like we belong.

Our only choice becomes blending in as much as possible - participating in activities we don't like: smiling politely when men make sexist comments 'as a joke, c'mon be a sport', not asking for a raise lest we be considered to be too pushy, drinking along even though we might get accosted or assaulted by a coworker, blending in as much as possible by wearing the same clothes, playing video games, ignoring casual groping at the office.

Adaptation and Anger

Sometimes, we are so successful with our adaptation that we are oblivious to the changes we had to make to achieve this success. We wonder why other women don't do the same? Why don't they try to remove their Otherness? It is not that hard! Look, I have done it!

This means sometimes we blame women who stand out as The Other. We tend to take a dim view of their discomfort and even more so, their anger. What fascinates me is that a lot of intellectual white males frequently tout the Roman/Greek Empire and their consequent violent conquests as inspirations for work ethics and startup success - but they are the very folks who rush to mark anger from us as 'violent' and hence not worth paying attention to. They bemoan the lack of nuance in understanding 'tech culture' from people who do not work in the tech industry, but are quick to make reductive statements about the world around us - especially when anger is expressed by people who feel uncomfortable in the field.

Ironically, they respond with sharp and violent reactions, which validates the anger that was originally expressed. The reactions vary from tone-policing to comments like: 'You have no more right to be angry than I have - I was nurtured in a broken home and bullied as a kid, if I can talk like a sensible person you can too.'

Some of The Others who have become sufficiently successful in this culture also feel uncomfortable with the anger. This discomfort expresses itself in many ways: Silencing the anger of The Others, or endorsing the status quo as it exists and expecting 'better' of The Others who feel uncomfortable. They are skeptical that this anger is justified.

Some of them are afraid this anger is going to endanger their success. There is fear that the dominant culture will decide it is no longer appropriate to promote women because they may *all* be this angry. Some of them do their best to actively combat this anger.

The Uses of Anger

But without this anger, those who conform or do not feel like The Other do not know about how vastly biased the workplace is. The anger makes us question why we conform to the workplace to begin with - was it because we were afraid of the consequences, because it appeared so natural? It also makes us wonder if we ourselves are one of The Others, who has done their utmost best to fit in.

Anger makes everyone aware of the urgency of the distorted comfort zones in the workplace. Some of The Others have opportunities to find other work environments to fit in, but if you are a woman in tech, chances of finding something you can fit into is fairly low.

Many times, it is Anger itself that gives space in the industry for other forms of contrary opinions to exist. Without Anger, The Others who were afraid of voicing their discomfort might not even speak out at all. Anger is what led to the outpouring of sexual assault incidents on Twitter using the #YesAllWomen hashtag.

Anger also leads to companies and people not wanting to become the cause of anger expressed by The Other. It has led to questioning of current processes, and consistent conversations around systemic biases in workplace culture. At the very least, it has kept discrimination in the front and center of conversations about tech culture.

And Anger also makes it imperative to take immediate actions.

The worst part of status quo in the workplace is how compelled everyone feels to stick to it. Until recently, conversations did not revolve around the discomfort and bias against The Others, but only on behaviours that The Others should adopt to become "Team Players". Several experts are readily available to tell us what to do ("change your name", "Cut your hair", "wear clothes that everybody else does", "speak with a different accent") to make sure the team likes us.

Our anger has blown that cover. We can now talk about systemic causes of discrimination and how The Others are driven out of workplace. Statistics from all the big tech companies prove that this is indeed the case.

What we need is *more* anger from those who are discriminated against, not less. We still do not have enough people talking vocally about discrimination they face for fear of recriminations. And until we have that, the biases we face in the workplace will not be corrected for.

Gender, Privacy and the Internet

Muna Mire

Kimberly Chiles comes across as both restrained and confident on the phone -- she's warm even while relaying a chilling experience.

"The thing that kept going through my head was: I'm almost 40, I'm a professional -- you know, this confident, put-together person and I worked really hard on having my shit together and basically, this [was] debilitating me," she tells me of her ordeal.

"Yeah."

"Fourteen-year-old me would not have been able to deal with it, I would have killed myself. There's not a question in my mind about that," she finishes quietly.

I don't know how to respond, so I wait for her to say something.

She tells me she wants to share her story with me, and that she feels it's important it gets out there. It's an ugly one -- after a bad breakup, an ex partner had her listed on a revenge porn website called MyEx.com. Her name and photo were listed along with obscene commentary -- and a link to her real Facebook profile. The subsequent harassment she endured was disgusting, but while the photos on the site were nonconsensually distributed, they were not of an intimate nature. When Chiles sought help from police, she tells me they politely informed her that she shouldn't have posted the photos to her Facebook in the first place and that there was nothing they could do because a crime had not been committed. There was no law on the books.

Chiles and I spoke recently after a mutual colleague connected us -- I was pursuing an assignment on anti-cyberbullying legislation, or rather "lawful access" legislation that included cyberbullying in its broad, unduly invasive scope. Chiles, along with the friend who connected us, feminist advocate Stephanie Guthrie, were called to testify before a committee convened in the Canadian parliament on their own experiences with what is alternately called "cyberbullying" or "revenge porn" in conversations around privacy and the internet.

Chiles and Guthrie were giving expert testimony on a bill that grants internet service providers (ISPs) complete immunity from all civil and criminal liability when handing over your private user data to the police, "peace officers" or "privacy officers," all without a warrant or court oversight given one is acting in the service of "administering of any law in Canada." A surveillance measure that basically sounds like it's straight out of 1984.

The legislation was proposed in the wake of the tragic, highly public suicides of Rehtaeh Parsons and Amanda Todd -- young women who were victims of what wider political discourses have termed cyberbullying. Proponents argue that the Internet and the way we have cultivated, defined, and networked it into the "real world" played as much of a role in what happened to Todd and Parsons as the men who assaulted them. Patriarchy thrives in the spaces online we have made so safe for perpetrators and so unsafe for girls, women, and non-male gendered people. The drive to enact a law to change the climate of online space is admirable but ultimately, without substantive discourses that centre the experiences of those most marginalized online, hugely problematic.

Chiles and Guthrie were very surprised to find themselves arguing passionately against legislation that was ostensibly enacted in their names as survivors of online harassment. They were both stunned that the law they had been advocating for for so long was an utter disappointment to them and their communities. Carol Todd, Amanda Todd's mother, also argued fiercely against the proposed legislation and the way it justified nonconsensual data probes and surveillance of online activity without a warrant. In spite of everything, Todd, who lost her daughter to cyberbullying and online harassment, believes that this sort of law is not a solution.

Guthrie argued before the legislature that "cyberbullying" is a term which perhaps fails to capture the gravity of what happened and continues to happen to young women like Parsons and Todd. She told me she dislikes the term "revenge porn" which is used more colloquially because in her view, pornography is ostensibly *consensually* produced. Instead, she uses the term "cyber-sexual assault" to describe what is, disturbingly, increasingly common behaviour online. Intimate photos -- or photo evidence of violent sexual assaults, trophies really -- are nonconsensually distributed online and shared by any number of perpetrators who then harass and mock victims, sometimes to death.

The term "bullying" evokes a schoolyard -- Guthrie reminds us that Parsons and Todd took their lives after being tormented in real time and then again, brutally, online. There is something truly sinister about the way that the violence enacted on their bodies was replicated ad infinitum in online space. The digital threads of Todd's story in particular serve as an exquisitely painful reminder and testament to her pain, witnessed and tacitly endorsed by those around her and critically, the structures (or lack thereof) we have in place governing online conduct.

Even more recently, the rape of a young woman named Jada and the online harassment that followed her was the stomach-turning subject of major news stories. Jada, just sixteen years old, was not defeated when photos of her assault went viral, as Chiles was so sure she would have been at her age. She decided to fight back with a social media campaign of her own, standing up to her rapists and those who mocked her rape online.

Sometimes, as in Chiles' case, even non-intimate photos are enough to cause serious harm, most often to women. For her, the damage came to her professional reputation since the offending material came up when you Googled her name. Chiles tells me that she could not step out of her house for fear of not knowing whether each person she encountered in her personal and professional life had access to the nonconsensually reproduced images. Who had access to the obscene material? Who had seen the results pop up in connection with her name on search engines? Who had access to her Facebook account -- the link having been posted on the revenge site?

She said she felt watched -- she felt surveilled.

Surveillance, Consent and Resistance

The spread of online violence -- whether replicating real time brutality against young girls and women via the digital image or not -- is a growing concern in conversations around privacy, surveillance and online space. Apart from cyber-sexual assault within the parameters of non-consensual image sharing, women are sent threatening messages, harassed and stalked online via dating websites -- or even on social media if they are particularly outspoken or have any semblance of a public profile. Women are most affected, racialized women even more so. In Canada, where the Parsons and Todd cases captured national attention, this gendered violence and the fear it rightfully garners has been used to justify legislation that encourages mass surveillance on the part of government and other state entities.

The sad irony lies in the fact that women like Chiles and Guthrie, young girls like Jada, Todd, or Parsons, all reported that they felt a similar panopticon effect of being watched or surveilled not from the government but from their assaulters and the wider community. The law is proposing to fix this by granting itself powers to surveil everyone -- including perpetrators. But is that too high of a price to pay to keep online spaces safe? More importantly, is that the cultural conversation we want to be having around privacy? One that, critically, steamrolls over the concept of informed consent?

The state, in the form of the legislature, decided that the best way to find and stop people who distribute pornographic images of minors, harass women, and generally distribute intimate images nonconsensually online was to push for a law that would allow anyone from the police to the mayor with a passing interest full access to your private user data. Wikipedia defines surveillance as "the monitoring of the behavior, activities, or other changing information, usually of people for the purpose of influencing, managing, directing, or protecting them." But as Chiles and Guthrie remind us, the key piece missing from this definition and more broadly from the wider discussion around privacy and online space is consent.

Parallels between surveillance as it is constituted by the state and the feeling of one's immediate community either tacitly witnessing or actively participating in what is often a gendered spectacle (e.g. cyber-sexual assault) exist. Critics of the law ask: Is there really a difference between teenaged boys accessing and sharing photographic evidence of a violent sexual assault online and the government accessing and sharing private user data? I submit that fundamentally, yes, there is a difference but that both actions are based on the same base violation of informed consent. Consent is never transferrable -- online, or in real time.

In wider conversations around surveillance, privacy and civil liberties, tacit consent to surveillance structures is assumed on the part of lawmakers representing citizens who enter into a social agreement with the government. Broadly speaking, in order to gain the state's protections as citizens, we agree to give up certain rights so as to protect the rest of our rights. It's part of the Hobbesian social contract -- and part of the state's origin story. The burden of proof lies with the citizen to speak out about this contract if it no longer serves the interests of the people -- e.g. when Edward Snowden revealed that the NSA conducts a global surveillance machine without public knowledge or consent. The way it is framed in popular political discourse, it's assumed that silence can act as tacit consent to state surveillance, and the moment one speaks out -- often at great personal risk -- is the moment when consent is revoked.

But imagine if we were to apply this conceptualization to sexual assault, which Todd, Parsons and Jada demonstrate is increasingly a concern online as well as in real time? Silence doesn't mean yes in real life! Consent is nontransferable -- agreeing to one act in a specific context doesn't mean you agree to other acts in other contexts. Arguably, we are no longer fighting that particular ideological battle. So why should it mean yes online? Why does the law treat perpetrators as though it does? Why, when a law is proposed to change this fact, does the legislation in fact violate informed consent on a mass scale? How can we possibly work to counteract digital violations of consent with further violations of consent? Structurally, it isn't a sound logic. And it further silences the needs of those who are ostensibly being helped.

In June, the Supreme Court of Canada agreed, placing an important condition on surveilling private citizens: you need a warrant to do so. But what did that take? Academics and journalists alike panned the bill for *months*. Advocates and victims themselves had to come to the court and explain why this bill was not about helping them. It was pretty one-sided in the eyes of the media: it was a bad bill. Unfortunately, the majority in the legislature didn't think so -- they passed it over the loud objections of the minority, the wider public, and even high-profile victims of cyber-sexual assault and their families.

In her speech to the committee, Guthrie stated that for her, "informed consent is an integral part of privacy." Guthrie -- and myself as a journalist -- draw from the work of Dr. Ann Cavoukian -- former Ontario Privacy Commissioner and surveillance studies scholar. Cavoukian pioneered "Privacy by Design," a value-sensitive systems engineering approach which operates according to a principle called "data minimization." Data minimization simply means that in creating a system, you prioritize minimizing the amount of private information collected and also informed consent in digital life.

Guthrie cited Cavoukian's explicit recommendations around both data minimization and "freely given and specific consent" as vital to centre in any discourses around state surveillance and digital privacy. Chiles and Guthrie know firsthand that legislation criminalizing online harassment and cyber-sexual assault is badly needed, and that the constellation of crimes proximal to cyber-sexual assault -- voyeurism, criminal harassment, extortion, blackmail -- are rarely enough to get the police to act on their own. Police inaction, the tendency to blame the victim (as in Chiles' case) rather than to take assault seriously -- in short, patriarchy -- ensures there is a gap between the law and its effective application. Why would police take cyber-sexual assault seriously when they don't take sexual assault seriously, period? Regardless, for Guthrie and for Chiles, the need for a law on the books to hold the police to account is necessary.

Ultimately, the state taking action against the myriad, deeply disturbing violence enacted in online space is necessary. The actions taken must begin and end with the voices of those most affected. If the state purports to be acting in their name it must centre their experiences in drafting legislation to keep online spaces safe. Were we to do that, we would note parallels between the patriarchal violence that exists in the so-called real world and the continuity online. We would also note the darkly ironic similarities between the actions taken by perpetrators of cyber-sexual assault and by the government in attempting to stop them. Violations of consent cannot be met with further blanket violations of consent on the part of the state. Online, offline, the song remains the same: we must centre the needs of survivors and use frameworks of consent culture to combat cyber-sexual assault.

Interview with Trans_ Anthology

Trans_ is the first anthology to collect the voices and experiences of trans people speaking to how the Internet has impacted their lives and how they have impacted the Internet. We sat down with editors Mitch Kellaway, nina de jesus and Harlan Kellaway to discuss the project, the publishing industry, and how to get involved.

Tell us about *Trans_*! What is it, what was the inspiration for it, and what are your goals?

Harlan: *Trans_* is the brainchild of Mitch and I. We're siblings who are both trans and we're very close - both as support for each other and as champions of each other's work. Mitch and I have independently gravitated towards doing work in the trans community related to our professions/passions, editing and tech respectively. We were texting one day as Mitch's last editing endeavor (*Manning Up: Transsexual Men on Finding Family, Brotherhood, and Themselves*) was winding down and I said something like "Wouldn't it be fun to create a book together?". We almost simultaneously came up with doing an anthology on trans people and tech; we chose the Internet given it impacts everyone's life in some way. We agreed it was an exciting topic and that our community undoubtedly had important and interesting experiences to share.

Lots of times, we see that the influence, impact and innovation of underrepresented and marginalized groups in technology gets erased. What do you hope to

explore or communicate about the roles, achievement, contributions and impact of trans people on the internet, in online activism and technology as a whole?

Harlan: The impact and innovation of underrepresented and marginalized groups in technology does get erased - and that certainly applies to trans people. But this project, I'd say, is less about the 'innovation' conversation and more about our community's more banal experiences with the Internet; what's unexceptional about underrepresented communities is often overlooked when we skip right to talking about innovation. What I hope to see in this work, in addition to an illustration of our community's achievements in tech and online activism, is a glimpse into the reality of trans people's lives as they intersect with the Internet.

There are lots of really amazing people behind *Trans_*! **Can you talk a little bit about the team and the perspectives and experience they bring to it? And how did you all meet and start working together?**

Mitch: I'm a transgender man who works as a freelance writer and editor. My brother Harlan works in web and app development and tech, and he had the original idea for an Internet anthology right around the time my first co-edited anthology on trans lives was released. We saw potential – albeit a pretty lopsided one – for a great book considering Harlan's passion and command around Internet issues, my own joy with the editing process, and our mutual investment in giving voice to trans people. But we immediately recognized that we needed to partner with a trans woman, and we considered approaching a number of accomplished women with experience in tech and editing.

I knew already of nina through a writing colleague, and was familiar with some of her exceptional work editing books by trans women with biyuti publishing. So we approached her and – I feel this is important to say – offered her fair compensation for our upcoming labor, and learned that she actually has some expertise with technology too. We all couldn't be a better fit in terms of skill-sets and interests, in my opinion. I can't wait until we're elbow-deep in a manuscript together.

nina: Yes! I had seen some of the call for papers going around and was/am considering submitting something myself since this is a nexus of issues/topics/whatever that I'm super passionate about.

Harlan: As opposed to nina and Mitch, I am not an editor. I studied and work in development. What I think is interesting about my background and journey here (besides studying Computer Science at a women's university as a trans man!) is that I had kept my identity as a trans person decoupled from my online presence for years - but I became inspired when I saw the call for the inaugural Trans*H4CK in September 2013. I decided I was done obscuring my trans identity, that I wanted to start talking publicly about what I do, about diversity in tech, and about tech projects for queer and trans people. And, without having gone through facing my fear of being out and subsequently being bolstered by a project that spoke so directly to me, I wouldn't have so readily suggested this project. Projects focused on trans people are so important!

I also think it's interesting that the production of this book is entirely facilitated by the Internet - Mitch and myself saw the work nina had put online, communicate online regularly, and have yet to meet outside of cyber-space.

You're working in this really interesting space of community organization, diversity, media and the internet. What do you see as the compelling things happening in independent publishing in diverse communities and

in technology in general? What other projects, organizations and developments inspire your team?

Mitch: The Internet changed the game for the publishing industry, and as far as I can tell it's been largely a positive thing for trans folks. Before print-on-demand technology and Internet-based self- and boutique publishing, trans writers or people producing books on trans topics had to approach the large publishing companies and try to make their work commercially "marketable" in the broadest sense – which means a lot of trans writing and characters (and especially those that complexified the simple narratives about us that appeal to "mainstream" cisgender audiences) never made it into print. I've heard much the same about writers and characters of color (both trans and cis).

The Internet has allowed small, independent presses to exist without the capital needed to print loads of books upfront; in the last few years, we've seen trans-focused presses like Homofactus Press, Topside Press, Trans-Genre Press, biyuti publishing, and Transgress Press pop up and produce some really fascinating, diverse work that would've never been possible otherwise.

Harlan: An interesting trend I've seen in tech, within the last half decade, is a big public push to get more women in tech. Big companies are making a big show to demonstrate they're invested in diversity and specifically gender diversity. I personally have misgivings about how deep that work goes; I wonder where trans people show up in that picture; and, more importantly, how maintaining the same backend to a system that has a problem with diversity is going to change how the system operates and what it outputs.

There is plenty of work already being done by the groups big companies are claiming to want to reach, and work that disrupts the notion these companies have frameworks we should aspire to. I'm talking projects like this publication - like the Fembot Collective, Trans*H4CK, the Empowermentors Collective, the critical and historical work by trans and cis people (anna anthropy, Mattie Brice, merritt kopas, Kortney Ziegler, Ashe Dryden, to name a few).

Tell us a bit about how you are going to be approaching the editing and publishing process.

Mitch: We've started with a call for nonfiction proposals only, and offered a bunch of potential topics. This anthology could cover a potentially endless number of topics, and we need to get a pulse on what people are thinking about and experiencing. We've also directly asked trans people whose writing we admire to consider contributing. We've had a huge amount of interest, and we'll definitely be able to start targeting themes with this first round of proposals. But we likely will not be able to capture every topic we'd like to hit – which often happens in a first round of submissions. So while we contact those whose proposals we've accepted, we may issue a second, more targeted call for essays.

Once we are able to articulate clearly our ideas of what the final product might possibly look like, we're going to fundraise for book promotion (something most presses ask you to do regardless of their size) and to offer our contributors a fair honorarium for their work. Then we will shop our book to publishers... we'll have to have a long conversation about where to submit it. I have publishers whose politics I love, and nina and I already both work for trans-focused presses – so it's exciting to say that we have strong reason to believe this book will come to fruition no matter what (though I'm excited to see which press it lands with). That means we can really focus on the craft rather than worrying about if it won't even see daylight because it's not "marketable" enough.

What advice do you have for others who are interested in working on their own media projects or publications in similar veins?

Mitch: I'll address the trans writers out there: write and submit, then write more and submit more. There are tons of literary journals that need trans input, and a publication history can support your future dreams of publishing a book (if that's something you aspire to). Self-publishing or starting your own LLC to print your work is totally an option because of the Internet – just make sure you have one or more editors. That is what separates professional self-published work that makes an impact on readers from the other 5,000+ books that are published each day. We, as trans writers and artists, have often missed out on being part of the professional writing world and the accompanying rigor and education that make excellent writers. Trans-focused presses are helping to turn that corner.

nina: As a person working within a small publishing company with a radically non-traditional approach to the entire editing/publishing process, I'd like to temper Mitch's advice about editors (even though, yes, I'm an editor), while editing is necessary... One of the great things about the current situation is that we also have many better opportunities to disrupt the notion that 'professional writing' must or can only be expressed within a narrow range of normative grammar and conventions. My suggestion is: write. Just write. Write as much as you can. Play around with style, form, and substance to figure out what works for you and, thus, makes your writing remarkable and spectacular.

Also? A good book designer and graphic designer makes a big difference for the overall 'quality' look of a book.

Harlan: I'd add that putting yourself out there - contacting organizations, using social media, talking to friends or groups that may support your work - is important.

Actually, most of the submissions we've received have come from people we haven't explicitly reached out to (though we have done a lot of that), they came from other people sharing the project, through social media campaigns, through other organizations posting our Call, etc; we wouldn't have their work if we were passive about outreach.

How can people get involved and support *Trans_*?

Harlan: If folks want to get involved past sending in proposals - they should keep an eye out for our crowdsourcing campaign. We already have a Patreon account set up for those who want to contribute a small amount regularly (www.patreon.com/trans_anthology). But, we will be doing a larger push later this year or early next year to gather the bulk of funds for a number of things - compensating writers, editors, and designers, marketing materials, events, etc.

If monetary means aren't an option - spread the word! We are aiming for publication in late 2015 and would love to see folks getting excited about and purchasing the book.

Just Focus On the Positives

Meagan Waller

I am a woman who works in the technology field. I am vocal about my experiences, passionate about creating a more empathetic infrastructure, and not a fan of sugar coating or coddling. When I speak out about my experiences as a woman in STEM, regardless of how optimistic or pessimistic my tone is, I'm met with the same tired comment from men with Really Good Intentions™:

"Maybe if you just focused on the positives, you'd realize that it's not so bad, it could be much worse."

Don't be mistaken though, this is rarely just friendly advice -- it's something much more ominous. At its core, asking victims and marginalized people to focus on the positives when they're talking about their lived experiences is a silencing tactic. It falls under derailment 101, and it does nothing to benefit people who are being oppressed. In fact, asking marginalized people to confront hate, oppression, and harassment with kindness and positivity itself holds oppressed people to a higher standard than that of their oppressors. This notion relies entirely on the sexist and racist expectations that marginalized people be deferential, polite, and in many cases grateful to their oppressors.

Being able to tune out the negatives and focus on the positives is a privilege that is usually only afforded to those who don't live on the margins. When I talk about my experiences as a woman, regardless of whether or not I bring up men at all, I get a few men asking me why I have to be so negative. The times that I do bring up men in my writing I get many messages, tweets and emails, all reminding me *"you can catch more files with honey than vinegar!"*

I don't know who decided this was a phrase that should be used to delegitimize righteous anger. I never have been, and never will be interested in catching or catering to pests, so can we just cut that out? The only purpose messages like that serve are to shift accountability onto the oppressed. In many instances it's to completely absolve oneself from responsibility or accountability altogether.

Telling a marginalized person that the way they break free of their oppression is through the power of positive thinking is possibly one of the oldest tricks in the book. If the experiences of marginalized folks existed in a vacuum, completely devoid of institutional power structures, maybe the power of positive thinking would be enough.

Nice = Good, Mean = Bad

There's this popular myth that has been perpetuated in media, culture, and religion since probably the dawn of time. That myth is that people who are nice are inherently good, and people who are mean are inherently bad. This myth holds up upon a really shallow examination, but once you start to scratch the surface you'll quickly see it crumble. How many neighbors of heinous criminals, mass murderers, serial killers, abusers, and rapists need to be interviewed saying, "We never expected it, they were such a nice person!" before we can finally move past this myth?

The most common tactic of an abuser is to always put on a nice front in front of other people. If they never act like an abuser in public it makes it much easier to gaslight, control, and abuse their victim. An article from RoleReboot.com by Soraya Chemaly, titled "How We

Teach Our Kids That Women are Liars", examines how media, politics, and religion perpetuate the message that women are not to be trusted. Since society already believes that women are liars by default, an abuser can simply take advantage of the myth that being nice means they're good and can operate with little to no suspicion, even when women speak out about their abusive behavior.

Our society has really internalized the notion that nice equals good and mean equals bad. I see it play out almost on a daily basis on Twitter, blog posts, and sometimes, on those rare occasions when I accidentally click a link taking me to Reddit or Hacker News. I see pseudo allies pitting feminists against one another. It's a sick game of good cop/bad cop, except the women in question have no say in their roles, and have no control over the narrative.

Women who are deemed too mean, too aggressive, too negative, still receive messages asking why they don't focus on the positives, but those questions get lost in the crowd of threats, harassment, and stalking from men who want to prove that her anger is invalid. Ironically, these techniques attempt to prove women are wrong or lying about the level of threats and gendered harassment they receive... by creating even more threats and gendered harassment. Conversely, women who are deemed acceptable, friendly, and approachable, are told constantly, "I'm so glad you're not like *her*", serving as a reminder that one misstep, tweet, blog post, or action deemed out of line can and will quickly be met with the same level of harassment.

"Real" Work

I've started to focus on accountability in my writing, in my actions, and in my conversations surrounding building empathetic and diverse communities in technology.

I've taught young women to code, mentored women, and volunteered with organizations working on eliminating the toxic culture surrounding our industry. Encouraging marginalized people through the pipeline is something that almost everyone, save for the most openly bigoted people, can agree on. It feels great to donate money, time, and resources to these organizations. It even feels great to simply advocate for these programs. However, this approach doesn't address any of the systemic reasons for the high attrition rates or harassment that marginalized people currently working in the field experience. This approach is but another way to defer accountability.

I've noticed a funny thing that has happened to me, and to other women in technology, when we talk about tech's toxic culture. We're met with men who have Really Good Intentions™, telling us:

"Instead of complaining about it online, why aren't you out in the REAL WORLD actually doing positive work for the community like @GirlsWhoCode?"

I always think this is pretty humorous, because I have taught Girls Who Code classes. I even met with members of the community, and executives from Girls Who Code, to discuss bringing their summer immersion program to Chicago. My experience isn't singular; many women have devoted their time, energy, and resources to address these issues only to have their contributions erased or invalidated for expressing any negativity about the field. In many cases, it's simply due to the fact that they're women.

When women start talking about how much it can suck to be a woman in technology, there is always at least one guy in her emails, mentions or comment section, letting everyone know that they *just can't agree* with her approach, but those programs that lead women into the pipeline, *that's real activism.*

That's something they can get behind.

Some of the most toxic people I've encountered have advocated for these programs and have condemned women who currently work in the field in the same breath.

What is it about these programs that make even some of the most toxic and terrible people in our field supporters? These pipeline programs allow people to feel warm and fuzzy inside with little to no effort. Advocating for pipeline programs requires no internal examination, just an outward one. You'd have to be deliberately obtuse to look at the state of our field and report back that white, cishet men are not in fact the majority. We can almost all agree that there is a lack of anyone who isn't part of that majority, and by advocating for these programs, we can *appear* to care... whether or not we actually do.

Many large companies with extremely homogenous teams donate large sums of money, host events, and work with these programs. These are initiatives that are seen as good, positive, and *acceptable*. These programs only require us to lead people to the pipeline. When, and if, they make it through the pipeline, they're treated as if it's *their* fault if they continue to be underrepresented and/or harassed to the point of leaving. These programs require absolutely no examination of the way we treat marginalized people who are already a part of the field.

Throwing more marginalized people into our industry without fixing the toxicity is not the answer. Asking for marginalized folks to meet oppression and hate with positivity, niceness, a smile, and silence are all great ways to ensure that you'll never have to think about the ways in which you actively, consciously, and subconsciously contribute to and reinforce the kyriarchy. You might even acknowledge that systematic oppression exists, but simply acknowledging it without actively examining yourself, educating yourself, and shutting up and listening instead of centering your hurt feelings does nothing to dismantle those oppressive structures.

The next time you feel like someone could achieve their goals more effectively if they could just be more positive, and less angry, stop. Ask yourself why you have no problem seeing the positive in the face of injustice.

Why aren't *you* angry?

Contributors

Julie Ann Horvath

A fan of the social web, building things, and making information beautiful, Julie Ann Horvath works as a designer, front-end developer, and visual storyteller at &yet. She was an early employee at GitHub, Cherry, and Yammer, and most recently founded Passion Projects; the industry's first all-woman talk series designed to surface and celebrate the work of incredible women in tech. She spends a lot of her time speaking at conferences about design, open source software, and people-first teams, as well as teaching workshops and volunteering to help (more) women and PoC learn to code.

Marco Rogers

Marco Rogers is Director of Frontend Engineering at Yammer, a part of Microsoft. He has been a part of the web development community for 10 years. These days his focus has shifted from programming to learning how to build a healthy and diverse engineering organization. He's also on Twitter occasionally.

Paul Riser, Jr.

Paul Riser, Jr. serves as managing director, technology-based entrepreneurship at TechTown Detroit, where he guides the design, execution and ongoing management of the nonprofit's technology programming: acceleration, proof of concept, university and health care system commercialization, and incubation. Riser is an entrepreneur, chief technologist, information officer and systems engineer with more than 17 years of professional experience working with start-up firms to global enterprises in the areas of IT systems administration, systems architecture, technology integration, technical presales support and medical device development.

Betsy Haibel

Betsy Haibel is a software engineer based in Washington DC. She's also a RailsGirls DC mentor, an organizer for Spanning Tree (DC's incipient feminist hackerspace), one-half of the feminist code collective Irregular Gentlewomen, and an occasional set designer. Her writing may be found at betsyhaibel.com.

Divya Manian

Divya works at Adobe on tools to help web designers. Previously, she worked at Opera, advocating the use of web technologies while contributing to open source projects and books on web development. She also participates in projects to make gender bias history like learn-to-search.com and aboutfeminism.me.

Muna Mire

Muna Mire is freelance journalist and a Black girl from the future. Her writing can be found at VICE.com and at Youngist.org, where she is an editor.

Harlan Kellaway

Harlan Kellaway is a trans application and web developer most interested in how tech and gender intersect, especially regarding trans communities. He has been involved in a number of projects including tech events for trans and gender nonconforming people, a platform to increase access to academic digital tools, and social media strategy for projects benefiting trans and queer people.

nina de jesus

nina is a digital projects librarian, blogs as satifice (satifice.com), and operates on the principle of good enough.

Mitch Kellaway

Mitch Kellaway is a trans writer, independent researcher, and co-editor of *Manning Up: Transsexual Men on Finding Brotherhood, Family and Themselves.* He is currently an assistant editor for Transgress Press and covers literature, family rights, and transgender issues for the Advocate.com. His other writing appeared in the *Lambda Literary Review, Original Plumbing, PolicyMic, Huffington Post, Zeteo Journal, Outrider Review*, and several queer and trans* anthologies.

Meagan Waller

Meagan is a software developer currently living in Chicago. She's enthusiastic about creating more inclusive and empathetic infrastructures in STEM fields. She has taught high school girls how to code; and is an advocate for mentoring women. Meagan loves to contribute to open source software; the project she's most involved in is exercism.io, a website providing crowd-sourced mentorship. She enjoys public speaking, and has recently started speaking at user groups and conferences. When she's not writing code she's on Twitter (@meaganewaller), grabbing a hoppy craft beer, or exploring Chicago.